STARTING OUT

I0896085

baby HAWKS

MEG GREVE

CREATIVE EDUCATION • CREATIVE PAPERBACKS

CONT

ENTS

I AM AN EYAS.

I am a baby hawk.

I am called an eyas (EYE-uhs).

See my feathers? They are soft and downy.

I hatched from an egg.

I will stay in the nest
until I learn to fly.
I need lots of rest.

My nest is high in a tree. My mom stays with me. My dad hunts for our food.

I call for food.
My mom spits it
into my mouth.

My flight feathers grow in.

After six weeks, I learn to fly. I use my talons. I can hunt for meat.

Now I can take care of myself!

SPEAK AND LISTEN

EEE!

Can you speak like an eyas?

Baby hawks shriek and squeak.

Listen to these sounds:

https://www.youtube.com/watch?v=1WSTelKRUkg

Now it is your turn!

EYAS WORDS

downy: soft and light

hatched: broke out of an egg to be born

hawk: a large bird that has a hooked beak and that hunts small animals

talons: sharp claws on a bird's feet

READING CORNER

Brereton, Catherine. *Nature Guide: Birds*. London: Bloomsbury Children's Books, 2020.

Riggs, Kate. *Hawks*. Mankato, Minn.: Creative Education/Creative Paperbacks, 2022.

Ripley Entertainment. *Ripley Readers: Raging Raptors*. Versailles, Ind.: Ripley Entertainment, Inc., 2021.

INDEX

PUBLISHED BY CREATIVE EDUCATION AND CREATIVE PAPERBACKS
P.O. Box 227, Mankato, Minnesota 56002
Creative Education and Creative Paperbacks are imprints of The Creative Company
www.thecreativecompany.us

LIBRARY OF CONGRESS CATALOGING-IN-PUBLICATION DATA
Names: Greve, Meg, author.
Title: Baby hawks / Meg Greve.
Description: Mankato, Minnesota : Creative Education and Creative Paperbacks, [2025] | Series: Starting out | Includes bibliographical references and index. | Audience: Ages 4-7 | Audience: Grades K-1 | Summary: "Introduce beginning readers to the soaring world of baby hawks with this life science starter. Includes photos, a labeled animal diagram, "Make a Noise" section, glossary, and further resources"-- Provided by publisher.
Identifiers: LCCN 2024014684 (print) | LCCN 2024014685 (ebook) | ISBN 9798889892076 (library binding) | ISBN 9781682775721 (paperback) | ISBN 9798889893172 (ebook)
Subjects: LCSH: Hawks--Juvenile literature. | Hawks--Infancy--Juvenile literature.
Classification: LCC QL696.F32 G748 2025 (print) | LCC QL696.F32 (ebook) | DDC 598.9/441392--dc23/eng/20240419
LC record available at https://lccn.loc.gov/2024014684
LC ebook record available at https://lccn.loc.gov/2024014685

DESIGN AND PRODUCTION
Design by Rhea Magaro
Production by Beeline Media and Design, Inc.
Art direction by Tom Morgan

PHOTOGRAPHS by Alamy Stock Photo/Gerrit Vyn, 9, Natural Selection David Ponton, 7, PAUL R. STERRY, 11; Dreamstime/Empire331, 4; Shutterstock/2009fotofriends, 10, Andrei.B, 14, Bohbeh, 13, Breck P. Kent, 8, Carol Gray, 12, Eric Isselee, cover, 2, Tom Meaker, 5, Vishnevskiy Vasily, 6

Printed in China